Rocks

Illustrations: Janet Moneymaker
Design/Editing: Marjie Bassler

Rocks
ISBN 978-1-950415-30-4

Published by Gravitas Publications Inc.
Imprint: Real Science-4-Kids
www.gravitaspublications.com
www.realscience4kids.com

If you walk outside and start digging with a shovel, you will discover rocks!

Rocks are everywhere!

THUNK!

All rocks come from **magma** deep inside the Earth.

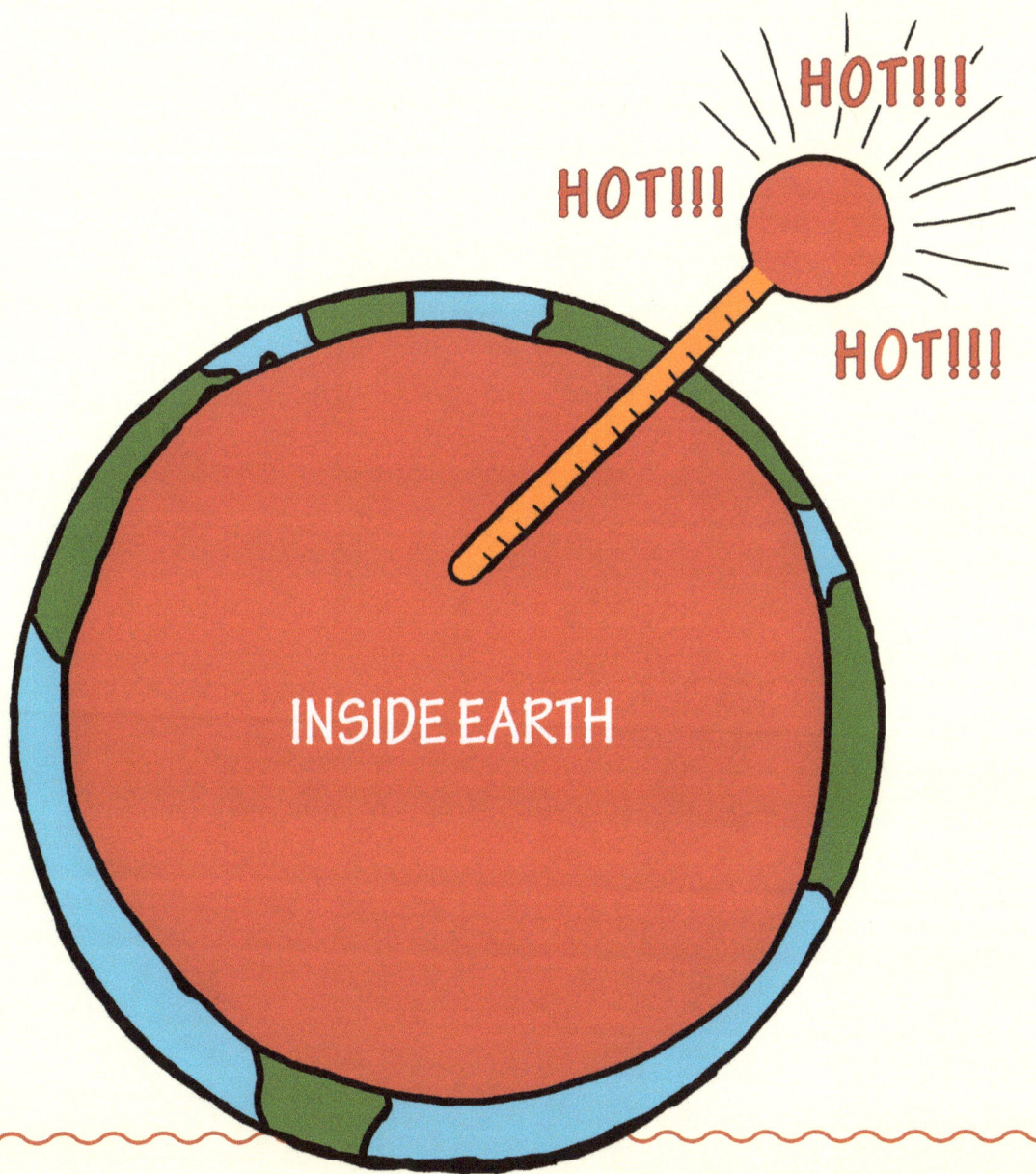

HOT!!! HOT!!! HOT!!!

INSIDE EARTH

Magma is very hot, **molten** (melted) material. Rocks form when magma cools.

Volcanoes happen when magma comes out from the inside of Earth.

Rocks from cooled magma

Magma

There are three main types of rock. These are **igneous, sedimentary,** and **metamorphic.**

Igneous
(IG-nee-us)

Sedimentary
(sed-uh-MEN-tuh-ree)

Photo by Moondigger,
CC BY SA 2.5

Metamorphic
(met-uh-MAWR-fik)

Photo by Huhulenik,
CC BY SA 3.0

(Examples)

Igneous rocks are made from magma that has cooled inside Earth.

Granite is a type of igneous rock.

Sedimentary rocks are formed from bits of rock left behind by wind or water.

Sedimentary rocks often have layers that can be seen.

Sandstone is a type of sedimentary rock.

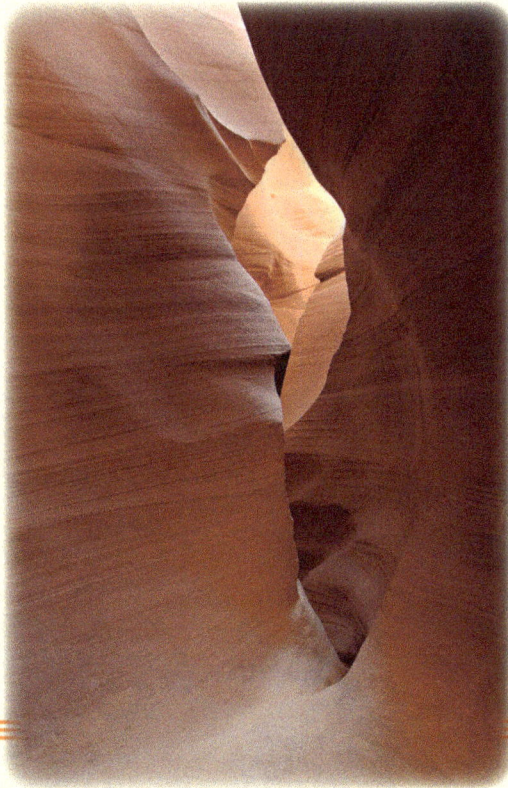

Metamorphic rocks are formed from other rocks that have been changed by heat and by being squeezed within the Earth.

Gneiss is a type of metamorphic rock.

How can geologists tell the difference between the three types of rocks?

I see these are different.

Geologists look at the features of rocks and do tests.

Geologists use chemistry too!

Learning more about geology will help you recognize the different types of rocks and how they form.

I want to learn more about geology!

How to say science words

geologist   (jee-AH-luh-jist)

geology   (jee-AH-luh-jee)

gneiss   (NIYSS)

granite   (GRAA-nuht)

igneous   (IG-nee-us)

magma   (MAG-muh)

metamorphic   (met-uh-MAWR-fik)

molten   (MOHL-tuhn)

sedimentary   (sed-uh-MEN-tuh-ree)

volcano   (vahl-KAY-noh)

What questions do you have about ROCKS?

Learn More Real Science!

**Complete science curricula
from Real Science-4-Kids**

Focus On Series

Unit study for elementary and middle school levels

Chemistry
Biology
Physics
Geology
Astronomy

Exploring Science Series

Graded series for levels K–8. Each book contains 4 chapters of:

Chemistry
Biology
Physics
Geology
Astronomy

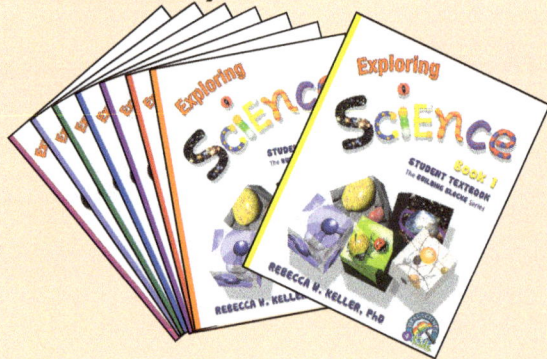

www.ingramcontent.com/pod-product-compliance
Lightning Source LLC
Chambersburg PA
CBHW040152200326
41520CB00028B/7574

9 781950 415304